YOUR KNOWLEDGE HAS VALUE

- We will publish your bachelor's and master's thesis, essays and papers

- Your own eBook and book - sold worldwide in all relevant shops

- Earn money with each sale

Upload your text at www.GRIN.com and publish for free

Sidanand Kambhar

Diversity of tree species in Gadag district, Karnataka, India

GRIN Publishing

Bibliographic information published by the German National Library:

The German National Library lists this publication in the National Bibliography; detailed bibliographic data are available on the Internet at http://dnb.dnb.de .

Imprint:

Copyright © 2014 GRIN Verlag GmbH
Print and binding: Books on Demand GmbH, Norderstedt Germany
ISBN: 978-3-656-83079-5

This book at GRIN:

http://www.grin.com/en/e-book/283046/diversity-of-tree-species-in-gadag-district-karnataka-india

GRIN - Your knowledge has value

Since its foundation in 1998, GRIN has specialized in publishing academic texts by students, college teachers and other academics as e-book and printed book. The website www.grin.com is an ideal platform for presenting term papers, final papers, scientific essays, dissertations and specialist books.

Visit us on the internet:

http://www.grin.com/

http://www.facebook.com/grincom

http://www.twitter.com/grin_com

DIVERSITY OF TREE SPECIES IN GADAG DISTRICT, KARNATAKA, INDIA

SIDANAND V. KAMBHAR*[1,2] & K. KOTRESHA[1]

[1]Floristic and Taxonomic Laboratory, Department of Botany, Karnatak University's, Karnatak Science College, Dharwad, Karnataka- 580 001, India.

[2]Post Graduate Department of Studies in Botany, Karnataka State Women's University, Jnanashakti, Torvi campus, Bijapur-586 109, Karnataka

ABSTRACT

The present work aimed to study the tree species diversity in Gadag district, Karnataka. A total of 133 tree species belonging to 105 genera and 42 families were encountered. Fabaceae was represented by the highest number of species (34), followed by Bignoniaceae (10). Among total number of the species 16 plants recorded as alien and medicinal plants respectively, beside this edible fruit 22 (including minor 16 and major edible fruit 6), dye yielding 7, fodder 4, sacred plants 5, timber 21 also been documented. It is essential to document tree species diversity, in order to gain more knowledge on species richness as well as their geographical distribution.

KEYWORDS: Gadag, Karnataka, Tree diversity

1

INTRODUCTION

Trees are the most important constituent of forests. It plays a major role in elucidating the patterns of distribution of biodiversity. These distributions are governed by biotic and a biotic factors (Ghate *et al.*, 1998; Upadhyay & Upadhyay, 2012). According to forester's and ecologist's a tree is defined as a woody plant that reaches diameter of 10cm (30cm girth) or more at the breast height (130cm above ground) (Chuyong *et al.*, 2011).

Trees provide basic requirements of human beings in the form of air, food, timber, plywood, paper, fuel wood, medicine and also give aesthetic value. The trees populations are disappearing at alarming rate due to deforestation, urbanization (Kishor *et al.*, 2011) and other various human needs (Ihenyen et al., 2009). In view of this fact, the present study is meant to prepare the checklist of tree species of Gadag district, the first exploration of the kind in this area.

MATERIAL AND METHODS

Study area:

Gadag district is geographically situated between latitudes 15° 16' and 16° 15' North and longitudes 75° 10' and 75° 55' East and it covers an area of about 4657 km². The district is divided into five taluka namely, Gadag, Mundargi, Nargund, Ron and Shirhatti. The distribution of the forests in the district is generally scattered and found in patches. These forests are comparable to Southern Thorn Forests of Champion and Seth's classification (1968). The main geographical feature of the district is the Kappat hills, a range of hills, with an elevation ranging between 300 and 1000m. The mean minimum temperature is 19°C during December to January and the mean maximum temperature is 42°C in May. The annual rainfall is generally less than 750 mm. The soil of Gadag district passes through every grade from bare rock to fairly deep loam with a thin covering of humus.

Floristic survey:

An extensive floristic survey was conducted during the year 2007-2011. The collected specimens were identified taxonomically with the aid of floras (Hooker, 1872-97; Cooke, 1958; Talbot, 1909 &1911, Saldanha, 1984 & 1996; Singh, 1988). and also Flora of India series(Sharma *et al.*, 1993; Sharma & Sanjappa, 1993; Hajra *et al.*, 1995a, b; Hajra *et al.*, 1997; Singh *et al.*, 2000). In the present study, arrangement of families was followed by APG III system of classification. The collected specimens were pressed and prepared herbarium followed by dry method of Jain and Rao (1977). The specimens were deposited in the Herbarium of Botany department, Karnatak Science College, Dharwad.

RESULTS AND DISCUSSION

The present article records 133 trees, which pertain to 105 genera of 42 families of flowering plants. Checklist of tree wealth is shown along with local name, category (cultivated, wild and alien), uses (edible fruits (minor and major), dye yielding, medicinal, fodder, sacred plants and timber) and collector number in Table 1.

Of the 42 families, 20 families are represented by single species and 8 families with 2 species each; 3 families are represented by 3 species each. Fabaceae is the dominant family with 34 species, followed by Bignoniaceae (10), Annonaceae (6), Arecaceae (6), Malvaceae (6), Moraceae (6), Anacardiaceae (5), Combretaceae (5), Myrtaceae (5), Apocynaceae (4). Of the total 132 species, 64 trees were recorded as cultivated form and remaining 69 species found in wild. Consequently, 16 species were documented as alien species for the district (Kambhar & Kotresha, 2011). Furthermore, 16 trees were recorded as medicinal plant species (Kotresha & Kambhar, 2010; Harihar & Kotresha, 2010; Kotresha & Harihar, 2011; Harihar & Kotresha, 2012). In addition to this, edible fruit 22 (including minor 16 and major edible fruit 6), dye yielding 7, fodder 4, sacred plants 5, timber 21 also been documented.

CONCLUSION

There are 133 trees species belongs to 105 genera and 42 families recorded for the Gadag District, Karnataka. Among total, 16 tree species were found as alien species, which have acclimatized in the district, while the rest 117 species are indigenous. Some indigenous species are becoming vanished due to anthropogenic pressure and urbanization. The old trees of the town as well as forest need to be protected as they provide glimpse of indigenous flora and a good habitat to several animals, bird species and many other associated species on them.

ACKNOWLEDGEMENTS

The authors are thankful to the University Grants Commission, New Delhi for providing financial assistance.

REFERENCES

CHAMPION, H.G. & S.K. SETH 1968. *A Revised Survey of the Forest Types of India*. New Delhi.

CHUYONG, G.B., D. KENFACK, K.E. HARMS, D.W. THOMAS, R. CONDIT & L.S. COMITA 2011. Habitat specificity and diversity of tree species in an African wet tropical forest. Plant *Ecology* 212: 1363-1374.

COOKE, T. 1958. *The Flora of Presidency of Bombay*. Vol. I-III, Bisen Singh Mehandra Pal Singh, Dehra Dun.

GHATE, U., N.V. JOSHI, M. GADGIL 1998. On the patterns of tree diversity in the Western Ghats of India. *Current Science* 75(6): 594-603.

HAJRA, P.K., V.J. NAIR & P. DANIEL 1997. *Flora of India*. Vol. 4, Botanical Survey of India, Culcutta.

HAJRA, P.K., R.R. RAO, D.K. SINGH & B.P. UNIYAL 1995a. *Flora of India*. Vol. 12, Botanical Survey of India, Culcutta.

HAJRA, P.K., R.R. RAO, D.K. SINGH & B.P. UNIYAL 1995b. *Flora of India*. Vol. 13, Botanical Survey of India, Culcutta. 1995.

HARIHAR, N.S. & K. KOTRESHA 2010. Wild Medicinal Plants of Kappat hills, Gadag District, Karnataka. *Research and Reviews in Biomedicine and Biotechnology* 1(2): 111-118.

HARIHAR, N.S. & K. KOTRESHA 2012. Wild Medicinal Plants of Kappat hills, Gadag District, Karnataka Part-II. *Life sciences Leaflets* 5: 37-42.

HOOKER, J.D. 1872-97. *Flora of British India*. Vol. 1-7. L. Reeve and Co., London.

IHENYEN, J., E.E. OKOEGWALE & J.K. MENSAH 2009. Composition of Tree Species in Ehor Forest Reserve, Edo State, Nigeria. *Nature and Science* 2009; 7(8): 8-18.

JAIN, S.K. & R.R. RAO 1977. *A handbook of field and Herbarium Methods*. Today and Tomorrow's Publishers, New Delhi.

KAMBHAR S.V. & K. KOTRESHA 2011. A study on alien flora of Gadag District, Karnataka, India. *Phytotaxa* 16: 52-62.

KISHOR, K., A.M. TRIPATHI, S. ROY & L.B. CHAUDHARY 2011. *Assessment and Preservation of Tree Diversity of Uttar Pradesh, India*. National Conference on Earth's Living Treasure Forest Biodiversity, Uttar Pradesh State Biodiversity Board. pp68-75.

KOTRESHA, K. & N.S. HARIHAR 2011. Uses of Cochlospermum religiosum (L) Alston Cochlospermaceae: An Ethno Medicinal Plant. *Indian Forester* 137(3): 393-394.

KOTRESHA, K. & S.V. KAMBHAR 2010. Traditional orthopedic treatment with medicinal plants. *Indian Forester* 136(9): 1281-1282.

SALDANHA, C.J. 1984. *Flora of Karnataka*. Vol. 1. Oxford and IBH Publishing Co. Pvt. Ltd., New Delhi.

SALDANHA, C.J. 1996. *Flora of Karnataka*. Vol. 2. Oxford and IBH Publishing Co. Pvt. Ltd., New Delhi.

SHARMA B.D., N.P. BALAKRISHNAN & M. SANJAPPA 1993. *Flora of India*. Vol. 2. Botanical Survey of India. Culcutta.

SHARMA B.D. & M. SANJAPPA 1993. *Flora of India*. Vol. 3. Botanical Survey of India. Culcutta.

SINGH N.P., J.N. VOHRA, P.K. HAJRA & D.K. SINGH 2000. Flora of India. Vol. 5. Olacaceae-Connaraceae, Botanical Survey of India, Culcutta.

SINGH, N.P. 1988. *Flora of Eastern Karnataka*. Vol. 1 and 2. Mittal Publications, New Delhi. 1988.

TALBOT, W.A. 1909. Forest Flora of the Bombay Presidency and Sind. Vol. 1. Government the Photozincographic Department, Poona.

TALBOT, W.A. 1911. Forest Flora of the Bombay Presidency and Sind. Vol. 2. Government the Photozincographic Department, Poona.

UPADHYAY R. & S.T. UPADHYAY 2012. Diversity of Trees at Hoshangabad, Madhya Pradesh. *Life sciences Leaflets* 2012; 6: 63-67.

Table 1. Checklist of tree wealth of Gadag district, Karnataka

Botanical name	Local name	Category	Uses	Coll. No.
ANACARDIACEAE				
Anacardium occidentale L.[+]	Godambi gida	C	EF	
Buchanania cochinchinensis (Lour.) Almeida[+]		W	EF	202
Lannea coromandelica (Houttuyn) Merill		W	F	389
Mangifera indica L.[++]	Mavin gida	C	EF, T	
Semecarpus anacardium L.f.[+]		W	EF,	610
ANNONACEAE		W		
Annona reticulata L.[+]		C, A	EF	
Annona muricata L.[+]		C	EF	
Annona squamosa L.[+]	Sitaphala	W	EF	353
Artabotrys hexapetalus (L.f.) Bhandari		C, A		
Miliusa tomentosa (Roxb.) Sinclair		W		781
Polyalthia longifolia (Sonnerat) Thwaites		C		
APOCYNACEAE				
Alstonia scholaris L.		C		
Plumeria alba L.		C		
Plumeria rubra L.	Kanigalu gida	C		
Wrightia tinctoria R. Br.	Halgathigida	W	D, M, T	205
ARALIACEAE				
Brassaia actionophylla Endl. var. *capitata* Clarke		C		
ARECACEAE				
Borassus fiabellifer L.		C		
Caryota urens L.		C		
Cocos nucifera L.[++]	Tengu	C	EF, M	
Livistona chinensis R. Br.		C		
Phoenix sylvestris (L.) Roxb.[+]	Ichala mara	W	EF,	289
Roystonea regia (H. B. and K.) O. F. Cook		C		

6

BIGNONIACEAE

Dolichondrone atrovirens (Heyne ex Roth) Sprague	Ured gida, Ulshin sappu	W	T	254
Jacranda acutifolia Humb. and Bonp.		C, A		
Kigelia americana (Lam.) Benth.		C, A		
Markhamia lutea (Benth.) K.Schum.		C		
Millingtonia hortensis L.f.	Akash mallige	C		
Spathodea companulata P. Beauv.		C, A		
Stereospermum chelenoides (L.f.) DC.		W		816
Tabebuia argentea (Bur. and Schum.) Britt.		C		
Tabebuia impetiginosa (Mart. ex DC.) Standl.		C		
Tabebuia pentaphylla Hemsl.		C, A		

BIXACEAE

Cochlospermum religiosum (L.) Alston	Bettadavare	W	M	799

BORAGINACEAE

Cordia dichotoma Forst. f.[+]	Kirjal gida	W	EF,	787
Cordia wallichi G. Don		W		961
Ehretia laevis Roxb.		W		769

BURSERACEAE

Boswellia serrata Roxb. ex Colebr.	Loban gida	W		775

CASURINACEAE

Casurina equisetifolia L.		C, A		

CELASTRACEAE

Cassine glauca (Rottlb.) Kuntz.		W		963

COMBRETACEAE

Anogeissus latifolia (Roxb. ex DC.) Wall. ex Guill. and Perr.	Dindala	W	T	6
Terminalia alata Heyne ex Roth	Budi balava	W	T	534
Terminalia catapa L.	Badami gida	C		
Terminalia cuneata Roth		W	D	381
Terminalia chebula Retz.	Alalekaigida	W	D, M, T	780

7

Cornaceae

Alangium salvifolium (L.f.) Wangerin ssp. *salvifolium*[+]	Ankaligida	W	EF, S	
EBENACEAE				
Diospyros melanoxylon Roxb. var. *melanoxylon*	Tumari gida	W	T	255
Diospyros montana Roxb.		W		257
EUPHORBIACEAE				
Givotia rottleriformis Griffith		W		813
Mallotus philippensis (Lam.) Muell.-Arg.	Kumkum mara	W	D,	776
FABACEAE				
Acacia auriculiformis A. Cunn		C	T	
Acacia chundra (Roxb. ex Rottl.) Willd.	Tered	W	M	286
Acacia leucophloea (Roxb.) Willd.	Belvantara, Belladagida	W	M	609
Acacia nilotica (L.) Willd. ex Del. ssp. *indica* (Benth.) Brenan	Kare jali	W	F, T	212
Adenanthera pavonia L.		C		
Albizia amara (Roxb.) Boivin	Tuggaligida	W	M	200
Bauhinia purpurea L.		C		
Bauhinia racemosa Lam.	Bangara mara	W		483
Butea monosperma (Lam.) Talbert	Muttaga	W		538
Caesalpinia coriaria (Jacq.) Willd.		C		
Cassia fistula L.	Kakkemara	W	D, M	
Cassia grandis L.f.		C		
Cassia javanica L.		C		
Cassia roxburghii DC.		C		
Dalbergia lanceolaria L.f. ssp. *lanceolaria*	Pachali	W	T	417
Dalbergia latifolia Roxb.		W		571
Dalbergia sissoo Roxb. ex DC.		C	T	
Delonix elata (L.) Gamble		C		
Delonix regia (Boj. ex Hook.f.) Raf.	Kempu	C, A		

	Gulmohar			
Erythrina suberosa Roxb.		W		822
Gliricidia sepium (Jacq.) Kunth ex Steud.	Gobbar gida	C	F	
Hardwickia binata Roxb.		W	F, T	535
Leucaena leucocephala (Lamk.) de Wit		C, A		
Parkia biglandulosa Wight and Arn.		C		
Parkinsonia aculeata L.		C, A		
Peltophorum pterocarpum (DC.) Backer and Heyne		C, A		
Pithecellobium dulce (Roxb.) Benth.		C, A		
Pongamia pinnata (L.) Pierre	Honge mara	W	M	258
Prosopis cineraria (L.) Druce	Bannimara	W	S, M	169
Pterocarpus marsupium Roxb.	Honne	W	T	611
Samanea saman (Jacq.) Merr.		C		
Senna siamea (Lam.) Irwin and Barneby		C		
Senna spectabilis (DC.) Irwin and Barneby		C		
Tamarindus indica L. [++]	Hunase mara	C, A	EF,	
LAMIACEAE				
Premna serratifolia L.		W		400
Tectona grandis L.f.	Sagvan gida	C	T	
LECYTHIDACEAE				
Barringtonia acutangula (L.) Gaertn.		W		292
Courouptia guianensis Aubl.	Shivling mara	C		
LOGANIACEAE				
Strychnos potatorum L.f.	Chillgida	W	M	721
LYTHRACEAE				
Lagerstromia indica L.		C, A		
Lagerstromia microcarpa Wight		W		808
MAGNOLIACEAE				
Michelia champaca L.	Sampige	C		
MALVACEAE				
Eriolaena quinquelocularis (Wight and		W		416

9

Arn.) Wight

Grewia orbiculata Rottl.	Uljeni gida	W		208
Grewia tillifolia Vahl[+]		W	EF	421
Guazuma ulmifolia Lam.		C		
Helicteres isora L.	Edumuri	W	M	39
Thespesia populnea (L.) Soland. ex Corr.	Bugari gida	C		
MELIACEAE				
Azadirachta indica A. Juss.	Bevinamara	C	M, S, T	
Melia azedarach L.		C, A	T	
Soymida febrifuga (Roxb.) A. Juss.	Sovi	W	D, T	692
MORACEAE				
Ficus benghalensis L.	Aladamara	C		
Ficus hispida L.f.		W		306
Ficus microcarpa L.f.		W		363
Ficus racemosa L.[+]	Atthimara	C	EF	
Ficus religiosa L.		C		
Streblus asper Lour.		W		791
MORINGACEAE				
Moringa oleiferu Lamk.[++]	Nuggi gida	C	EF	
MUNTINGIACEAE				
Muntingia calabura L.[+]	Sakare hannu	C, A	EF	
MYRTACEAE				
Callistemon speciosus (Sims) DC.		C		
Eucalyptus globulus Labill.	Neelgiri mara	C	M	
Psidium guajava L.[++]	Peral hannu	C	EF	
Syzygium cumini (L.) Skeels[+]	Kadu neral hannu	W	EF, T	431
Syzygium jambos (L.) Alston[++]	Niral hannu	C	EF	
OLEACEAE				
Schrebera swieteniodes Roxb.		W		840
PANDANACEAE				
Pandanus fascicularis L.		W		987
PHYLLANTHACEAE				

Bridelia crenulata Roxb.		W		413
Phyllanthus emblica L.[+]	Kadu nelli	W	EF, T	717
POACEAE				
Bambusa arundinacea (Retz.) Willd.		W		805
PROTEACEAE				
Grevillea robusta A. Cunn. ex R. Br.		C		
RUBIACEAE				
Gardenia latifolia Ait.	Adavi bikki	W		964
Mitragyna parvifolia (Roxb.) Korth.		W	T	247
Morinda pubescens J. E. Sm.	Malaga	W	D, M	388
RUTACEAE				
Aegle marmelos (L.) Corr.	Bilpatri, patri	W	M, S	282
Chloroxylon swietenia DC.	Masval gida	W	T	157
SALICACEAE				
Flacourtia indica (N. Burman) Merr.		W		693
SANTALACEAE				
Santalum album L.	Srigandha, Chandan gida	W		367
SAPINDACEAE				
Sapindus emarginatus Vahl		W		944
SAPOTACEAE				
Madhuca longifolia (Koen.) Macbride var. *latifolia* (Roxb.) Chevalier[+]	Hippali mara	W	EF	827
Manikara zapota (L.) Van Rayen[++]	Chiku gida	C	EF	
SIMAROUBACEAE				
Ailanthus excelsa Roxb.		W		523
STRELITZIACEAE				
Ravenala madagscariensis J.F. Garel.		C		
ULMACEAE				
Holoptelea integrifolia (Roxb.) Planch.		W		955
ZYGOPHYLLACEAE				
Balanites aegyptica (L.) Del.	Ingalarada gida	W	S	484

Category: C= Cultivated, **W**= Wild, **A**= Alien; **Uses**: **EF**= Edible Fruits, **D**= Dye Yielding, **F**= Fodder, **M**= Medicinal, **S**= Sacred plants, **T**= Timber, $^+$= Minor edible fruits, $^{++}$= Major edible fruits